Chemical Engineering & the Law Forum 2018

Core Programming Area at the 2018 AIChE Spring Meeting and 14th Global Congress on Process Safety

Orlando, Florida, USA
22 – 25 April 2018

ISBN: 978-1-5108-6451-1

Printed from e-media with permission by:

Curran Associates, Inc.
57 Morehouse Lane
Red Hook, NY 12571

Some format issues inherent in the e-media version may also appear in this print version.

Printed by Curran Associates, Inc. (2018)

For permission requests, please contact AIChE
at the address below.

AIChE
120 Wall Street, FL 23
New York, NY 10005-4020

Phone: (800) 242-4363
Fax: (203) 775-5177

www.aiche.org

Additional copies of this publication are available from:

Curran Associates, Inc.
57 Morehouse Lane
Red Hook, NY 12571 USA
Phone: 845-758-0400
Fax: 845-758-2633
Email: curran@proceedings.com
Web: www.proceedings.com

TABLE OF CONTENTS

Some abstracts in this book refer to images that were not made available by the publisher

513716 Justifying Energy Savings in Ethylene Crackers in a Low Energy Price Environment

Abstract:

Global energy prices have dropped to their lowest level in a decade and have taken with them the days of easy justification for energy improvements. In specific countries where energy prices are still subsidised, this is a familiar situation in everyday energy management. The subject of this paper is the different ways that have been used to improve the economics of energy projects in low price situations in order to make viable project investments.

Case studies will be used to illustrate such methods as:

- Debottlenecking energy critical equipment to increase throughput,
- Overcoming an energy cap to improve severity and product yield,
- Making energy savings to meet legislative targets, and
- Using energy reduction to justify changing catalysts to improve yield.

511653 Retrospective Health Impact Assessment for Pm 2.5 Pollution in Mexico City 1990-2015

Abstract:

Fine particulate matter has a well-documented association to mortality. We have conducted a retrospective health impact assessment to quantify the lives saved in Mexico City from 1990-2015 due to PM 2.5 reduction resulting from pollution control policies, utilizing the Integrated Exposure Response model developed for the Global Burden of Disease study. In addition, we monetized the health benefits using city-specific value of statistical life (VSL). Our analysis suggests that 21,000 lives saved during that period and benefits in the order of 18 billion 2015 USD. We also developed a novel model to retrospectively decompose the health benefits and attribute them to one policy utilizing sulfur reduction in fuel policy as an example. Our model indicates that this policy has resulted in 1,600 lives saved. This suggests that Mexico City air pollution policies could serve as a model for developing countries and the industry with environmental regulations at infancy stage.

512814 Sustainability Analytics - the Critical Link between Sustainability and Business Strategy for Actionable Insights

Abstract:

Advances in Enterprise Systems Are Making It Feasible for Corporations to Track and Transform Sustainability Performance. the Materiality of These Seemingly Non-Economic Impacts Is the Critical Link between Sustainability and Business Strategy. Business Analytics Enable Organizations to Convert Raw Data into Actionable Insights to Achieve Their Sustainability Goals. Sustainability Analytics Analyze the Data and Provide Insights That Help Transform Sustainability Information into Action across the Value Chain and Life Cycle. This paper will Provide Business Leaders with Insight into How to Determine Which Sustainability Metrics Are Material to Them and Relevant to Their Business, and How to Incorporate Them into Corporate Strategy, Resource Allocation Prioritization and Decision Making Process.

Based on My Book Introduction to Sustainability Analytics Under Publication By CRC Press

512839 The Reaction Mechanism and Kinetics of Magnesium Oxide for CO_2 Immobilization

Abstract:

The CO_2 immobilization is significant for the sustainable energy and chemistry industries. And the magnesium oxide is a clean and economic agent using for the CO_2 immobilization. Therefore, the reaction of CO_2 with magnesium oxide suspended in water was investigated, the mechanism proposed, and the kinetics regressed as well. The CO_2 pressures from 1 to 5 (MPa), as well as ratios of solid to liquid, stirring speeds and temperatures, significantly influenced the dissolution of magnesium oxide. And as high as 26.63 g/L MgO was dissolved in the solution under the condition of CO_2 pressure 5 MPa at 50 °C.

In the process, the magnesium oxide became magnesium hydroxide through hydration reaction, and then the hydroxide dissolved by carbonated water. In the carried out reaction of CO_2 with magnesium hydroxide suspension, no solid product was found covering the solid $Mg(OH)_2$. Therefore, a reaction mechanism based on the hydration of Mg^{2+} was proposed, where the concept of â€˜progressive liquid filmâ€™ was introduced to the reaction system. And the derived kinetic model for the dissolution was fitted the experiment data well. The pressurized magnesium hydroxide dissolution was found to be rate-controlled by the diffusion in the progressive liquid film around the solid material. The apparent activation energy, obtained from 17.1 to 25.4 (kJ/mol) in the experiments, indicated also that the dissolution inclined to be diffusion rate controlled. Moreover, the apparent diffusion coefficients of H^+ in the liquid film were calculated and an empirical expression was obtained for the first time. The kinetic model might be applicable to other dissolution reactions similarly with just soluble product.

KABOOM! RECENT MISADVENTURES IN THE PATENT WORLD

Gerard E. Reinhardt, Manhasset, NY

Biography

Gerard Reinhardt is an Intellectual Property attorney, and a principal in the Reinhardt IP law firm (see ReinhardtIP.com). Mr. Reinhardt is licensed to practice law in New York, Texas and the District of Columbia, and is registered to practice before the U.S. Patent & Trademark Office. In addition to a J.D. from the University of Houston, he holds B.S. and M.S. degrees in Chemical Engineering from UVa and Georgia Tech, respectively. He has litigated complex patent cases with the New York law firm of Morgan & Finnegan, and functioned for almost ten years as Senior Counsel to Schering-Plough Corp. and Merck & Co., acquiring an in-depth knowledge of practice in chemical and life science patents. He can be contacted at Gerard@ReinhardtIP.com.

Introduction

A patent can be analogized to a processing plant. Both are carefully designed, constructed and maintained at considerable expense. Both can play key roles in a business model. And both are subject to unexpected failures that can negatively impact (or even cripple) an enterprise.

This paper will examine the case of Allergan v. Apotex[1], in which a pair of chemical patents was found to be invalid, undoubtedly to the great surprise and discontent of the patent holders' management, inventors, lawyers and shareholders. Lessons from this case are generally applicable in avoiding the unpleasantries of the catastrophic failure of such a patent.

Background

Allergan, Inc. ("Allergan") has U.S. Food and Drug Administration ("FDA") approval to sell Latisse®, a 0.03% bimatoprost ophthalmic solution, as a topical solution to treat hypotrichosis (i.e., hair loss or reduction) of the eyelashes by stimulating hair growth. Bimatoprost is a synthetic prostaglandin F-2-alpha ("PGF") analog.
The chemical structure of bimatoprost is displayed below:

[1] Allergan, Inc. v. Apotex Inc., 754 F.3d 952, 2014 U.S. App. LEXIS 10710, 111 U.S.P.Q.2D (BNA) 1245, 2014 WL 2579287.

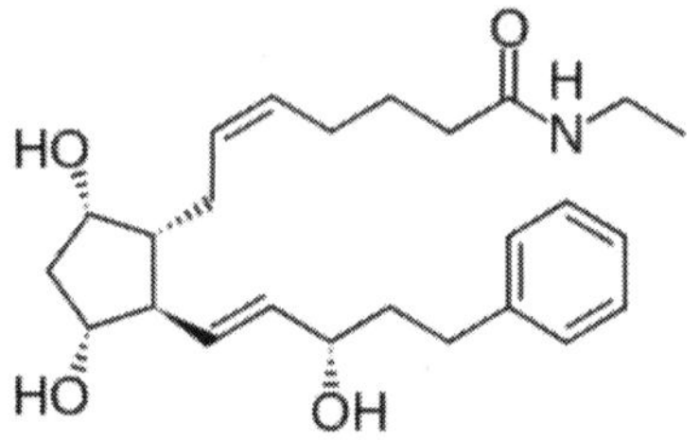

Bimatoprost chemical structure

Referring to the two chains of atoms attached to the cyclopentane, the top chain is called the "alpha chain," and the chain below it is called the "omega chain."

By the mid-1980s, it was established that naturally occurring PGF could alleviate intraocular pressure ("IOP"), which is associated with the eye disease glaucoma. To develop an effective treatment and minimize side effects, scientists worked on synthesizing and testing more selective derivatives and analogs of PGF. A particular kind of PGF analog, 17-phenyl PGF analogs, proved to be particularly useful in the treatment of glaucoma. Bimatoprost is one such synthetic 17-phenyl PGF analog, which emerged in the course of research by Allergan scientists. In 2001, Allergan received FDA approval to sell Lumigan®, a 0.03% bimatoprost ophthalmic solution—identical strength to Latisse®—as an eyedrop to treat glaucoma, which it continues to market.

In the 1990s, Dr. Murray Johnstone performed studies on the use of latanoprost, another 17-phenyl analog for treatment of hair loss. Latanoprost optical solution also received FDA approval for use in glaucoma treatment, and it continues to be marketed as Xalatan®. Dr. Johnstone observed that in the course of treating glaucoma patients with latanoprost eyedrops, a substantial fraction of them grew much longer and denser eyelash hair. Dr. Johnstone filed a patent application on the use of latanoprost and other 17-phenyl PGF analogs to promote hair growth in February 1997.

The work that led to the '029 patent, the first of the two patents asserted by appellees in this case, was conducted by researchers at Proctor & Gamble led by Dr. Mitchell DeLong. Dr. DeLong and his team studied the effects of a wide range of prostaglandin compounds in mice. In the course of their studies they observed that administration of PGF compounds that were selective for the FP receptor resulted in growth of longer and thicker hair. On March 21, 2000, Dr. DeLong and others filed a provisional patent application on the topical application of compounds that bind the FP receptors to treat hair loss. U. S. Patent no. 7,388,029 ("the '029 patent") claims priority to this provisional application. During prosecution, in 2003, the parent application of the '029 patent was assigned to Duke University, and the patent issued on June 17, 2008.

The second patent asserted in this suit is U.S. Patent no. 7,351,404 ("the '404 patent"), which is assigned to Allergan. The '404 patent arises from observations made during the clinical trials for Lumigan®. As had been observed for latanoprost, glaucoma patients treated with bimatoprost eyedrops spontaneously grew longer and thicker eyelash hair[2]. The '404 patent covers the treatment of eyelash hair loss through topical application of bimatoprost, and it claims priority to a provisional application filed on February 4, 2002.

Allergan and Duke University (referred to hereinafter as "Allergan") sued each of the appellants (hereinafter "Apotex")[3] after they submitted Abbreviated New Drug Applications ("ANDAs") to the FDA seeking to market a generic version of Allergan's Latisse® product. Allergan asserted claims 1, 8, 14, 18, and 20 of the '029 patent and claim 14 of the '404 patent. After a bench trial in the consolidated Hatch-Waxman action, the district court held, *inter alia*, that the asserted claims of the '029 and '404 patents are not invalid for anticipation, obviousness, insufficient written description, or lack of enablement, and, moreover, that Apotex infringed[4]. The district court subsequently enjoined Apotex from commercial manufacture, use, offer to sell and/or sale of the proposed products until the latest of the expiration dates of the '029 and '404 patents.

An appeal to the U.S. Court of Appeals for the Federal Circuit ("the Federal Circuit") followed, in which Apotex raised issues of claim construction for the '029 patent, as well as the invalidity of the asserted claims of the '029 and '404 patents.

I. Claim Construction of the '029 Patent

Apotex raised a single claim construction issue on appeal concerning the '029 patent. The '029 patent's asserted claims are directed towards a method of "treating hair loss." Apotex challenged the district court's construction of this term as meaning that the invention may arrest hair loss, reverse hair loss, or promote hair growth *in the alternative*.

The specification provides an express definition for the term: "'Treating hair loss' includes arresting hair loss or reversing hair loss, or both, and promoting hair growth."[5] Apotex argued, however, that use of the conjunctive "and" in the inventor's own lexicography expressly provides that the method for treating hair loss must *both* arrest or reverse hair loss, as well as *also* promote hair growth. Apotex argued that under their proposed construction, a generic version of Latisse® would not infringe because Latisse® treats hair loss by lengthening, thickening, and darkening existing healthy hair—which Apotex argued means only the promotion of hair growth.

The district court found that the use of the word "includes" in the definition of "treating hair loss" plainly means that the patentee intended to define treating hair loss to include the possibility of one or all of arresting hair loss, reversing hair loss, or promoting hair growth. Numerous examples in the patent describe the use of claimed compositions to "induce hair

---

[2] *See* '404 patent, col. 11 ll. 5-62.
[3] Under 35 U.S.C. § 271(e)(2)(A).
[4] *Allergan, Inc. v. Apotex, Inc.*, Nos. 1:10-CV-681, 1:11-CV-298, 1:11-CV-650, 2013 U.S. Dist. LEXIS 13987, 2013 WL 286251, at *1 (M.D.N.C. Jan. 24, 2013).
[5] '029 patent, col. 3 ll. 29-30.

growth," "darken and thicken eyelashes," "promote hair growth," and "promote eyelash growth."[6] The court found that there is nothing in either the specification or the claims suggesting that the patentee would have excluded these examples from the scope of claimed methods. The Federal Circuit agreed, thus upholding the trial court's finding of infringement of the '029 patent.

II. Invalidity by Anticipation of the '029 Patent

A patent is invalid for anticipation under 35 U.S.C. § 102[7] if a single prior art reference discloses each and every limitation of the claimed invention.[8] Anticipation is a question of fact reviewed for clear error.[9]

With respect to the '029 patent, Apotex raised two allegedly anticipatory references: (a) the published patent application arising from Dr. Johnstone's aforementioned research on promoting hair growth using latanoprost and related compounds, and, (b) Allergan's earlier patent on the use of bimatoprost and related compounds to treat glaucoma.

a. Johnstone PCT Application

Dr. Johnstone's research led to the filing of International Patent Application No. PCT/US98/02289 ("Johnstone") on February 3, 1998. In particular, Johnstone discloses methods for stimulating hair growth using a broad genus of prostaglandin analogs that features a substituted five-carbon alicyclic ring with one or more optional double bonds, an esterified six-carbon alpha chain including a double bond between the second and third carbons from the ring (C_5 and C_6), and optionally substituted terminal carbon, and an optionally substituted 3- to 16- carbon omega chain optionally featuring a double or triple bond between the second and third carbons from the ring. The combinations of moiety definitions disclosed in Johnstone cause this genus to encompass tens or perhaps hundreds of thousands of compounds. Johnstone further discloses the structures of 30 species of compounds within the genus. However, in one instance, Johnstone also discloses compounds with a single (also known as "saturated") bond at the C_5-C_6 position: "The chain could preferably be a C_6-C_{10} chain which can be *saturated or unsaturated*, having one or more double bonds, and allenes, or a triple bond."[10]

---

[6] *See, e.g.*, '029 patent at cols. 58 ll. 17-19, 59 ll. 43-44, 59 ll. 61-62, 60 ll. 31-32, 60 ll. 11-12.

[7] The America Invents Act ("AIA") took effect on September 16, 2012. Because the filing dates of the applications from which the patents at issue here granted are prior to that date, pre-AIA versions of 35 USC §§102 and 103 are applied in this case.

[8] *Schering Corp. v. Geneva Pharms., Inc.*, 339 F.3d 1373, 1377 (Fed. Cir. 2003).

[9] *Atlas Powder Co. v. Ireco, Inc.*, 190 F.3d 1342, 1346 (Fed. Cir. 1999).

[10] Johnstone at 12 ll. 22-24 (emphasis added).

During prosecution of the application that eventually granted as the '029 patent, the applicant disclosed U.S. Patent No. 6,262,105 to Johnstone, which shares its disclosure with the Johnstone PCT application. Following issue of the initial notice of allowance, the applicant filed an amendment to add a proviso to independent claims with the aim of expressly excluding the thirty species disclosed in Johnstone. The examiner accepted the amendment and issued a notice of allowance for the patent with the claims amended to include the provisos.

At trial, Apotex argued that Johnstone's single recitation describing alpha chains in which the C_5-C_6 bond is saturated anticipated the '029 patent. Johnstone, however, contains no other disclosure of a structure with a single, saturated, bond at that location. All of the examples of Johnstone are drawn to compositions in which the C_5-C_6 bond is a double bond.

The district court found that in the context of the state of the art, a person of ordinary skill would not read Johnstone as disclosing a single bond structure. The district court accepted Allergan's evidence that a PGF analog with such a structure would not have been thought to have a therapeutic effect because it would not selectively bind to the FP receptor.[11] The district court further supported its determination by citing to the '029 patent's notice of allowance, in which the examiner indicated Johnstone "lacks motivation to modify the prostaglandins taught therein in order to obtain the presently claimed prostaglandins."[12] The district court found, therefore, that the disclosure on which Apotex relied did not "clearly and unequivocally disclose" the use of compounds with the saturated bond.[13]

On appeal, the Federal Circuit was persuaded that in light of the very limited disclosure, the trial court did not commit clear error. The disclosure was too sparse and ambiguous for a person of ordinary skill to comprehend Johnstone's disclosure as anticipating the '029 patent. Accordingly, the Federal Circuit affirmed the district court's finding that the Johnstone did not anticipate the '029 patent.

b. Inherent Anticipation by the '819 Patent

U.S. Patent No. 5,688,819 ("'819 patent") emerged from research by Allergan scientists on selective PGF analogs that could effectively treat glaucoma. The '819 patent discloses the use of a set of selective PGF analogs, including bimatoprost, which is specifically identified by its chemical structure.[14] The '819 does not refer to hair growth or treating hair loss, nor does it disclose topical application of any compounds.

At trial, Apotex argued that because (i) the '819 patent's disclosure teaches the application of eyedrops containing compounds within the scope of the asserted '029 patent claims, in particular bimatoprost, and (ii) the application of eyedrops containing bimatoprost

---

[11] *Allergan*, 2013 U.S. Dist. LEXIS 13987, 2013 WL 286251, at *5.

[12] *Id.*; J.A. 5559-65 at 5564.

[13] *Allergan*, 2013 U.S. Dist. LEXIS 13987, 2013 WL 286251, at *6 (citing *Net MoneyIN, Inc. v. VeriSign, Inc.*, 545 F.3d 1359, 1371 (Fed. Cir. 2008)).

[14] '819 patent, col. 7 ll. 44-46.

results in the growth of eyelashes, the disclosed method inherently anticipates the '029 patent.

A single prior art reference may anticipate without disclosing a feature of the claimed invention if such feature is necessarily present, or inherent, in that reference.[15]

At issue was whether promoting hair growth through topical application of bimatoprost on the skin is necessarily present or inherent in the method of applying eyedrops containing bimatoprost. There was no dispute that the application of eyedrops containing bimatoprost *can* result in the promotion of eyelash hair. Allergan's own description of the conception of the '404 patent was that scientists first observed that after monkeys were treated with bimatoprost in their eyes, some of them showed enhanced eyelash growth. Similar results were found in clinical trials, during which some patients who used bimatoprost eyedrops reported eyelash growth.

The district court found that Allergan's expert witness, Dr. Noecker, had persuasively testified that a "properly applied drop" would not transfer to the skin.[16] The district court additionally found that Lumigan® clinical trials only showed that a fraction of patients experience eyelash growth, and held that because a bimatoprost eyedrop only *may* contact skin, the '819 patent does not inherently anticipate as inherency "may not be established by probabilities or possibilities."[17]

On appeal, Apotex argued that the district court erroneously required *certainty* as a prerequisite for inherent anticipation. In their view, the district court should not have relied on the finding that only some patients who received eyedrops experienced eyelash growth. The Federal Circuit was not persuaded that the district court committed clear error in finding that the '819 patent does not inherently anticipate the claims of the '029 patent, and affirmed.

III. Obviousness of the '029 Patent

Apotex alternatively alleged that the aforementioned references render obvious the asserted claims of the '029 patent. A patent is invalid for obviousness "if the differences between the subject matter sought to be patented and the prior art are such that the subject matter as a whole would have been obvious at the time the invention was made to a person having ordinary skill in the art to which said subject matter pertains."[18]

Obviousness is a legal conclusion based on underlying facts.[19] Underlying factual inquiries include (i) the scope and content of the prior art; (ii) the differences between the prior art and the claims at issue; (iii) the level of ordinary skill in the field of the invention; and (iv)

---

15 *Schering Corp. v. Geneva Pharms., Inc..*, 339 F.3d 1373, 1377 (Fed. Cir. 2003).

16 *Allergan*, 2013 U.S. Dist. LEXIS 13987, 2013 WL 286251, at *6.

17 *Id.* (citing *Bettcher Indus., Inc. v. Bunzl USA, Inc.*, 661 F.3d 629, 639 (Fed. Cir. 2011).

18 35 U.S.C. § 103(a).

19 *Graham v. John Deere Co.*, 383 U.S. 1, 17, 86 S. Ct. 684, 15 L. Ed. 2d 545 (1966).

relevant secondary considerations including commercial success, long-felt but unsolved needs, failure of others, and unexpected results.[20]

The district court held that the '029 patent was non-obvious, based principally on its finding that there was no motivation to combine Johnstone and the '819 patent due to "pharmacological differences" between the compounds that each reference discloses.[21] Specifically, the '819 patent discloses a list of thirteen chemical compositions for 17-phenyl PGF analogs, which were like those disclosed in Johnstone in that they contain a C_5-C_6 double bond and have demonstrably high pharmaceutical activity in treating intraocular pressure with minimal side effects.[22] However, the '819 patent's compounds contain a C_1-amide group whereas Johnstone generally discloses compounds with esters or carboxylic acids at the C_1 location. The district court was persuaded by Allergan's expert that it was understood at the time of the invention of the '029 patent that because of this C_1-amide group, bimatoprost and the other 17-phenyl PGF analog compositions disclosed in the '819 patent were thought to bind to a receptor other than the FP receptor.

The district court determined that secondary considerations additionally weighed in favor of non-obviousness, based on its finding that the invention of the '029 patent was an unexpected result (for the same reasons that it found a lack of reasonable expectation of success), as well as the commercial success of Latisse® and the paucity of competing hair growth treatments.

On appeal, the Federal Circuit observed that the district court reached its conclusion of nonobviousness by looking only at properties of the C_1-amide group and, particularly, bimatoprost, and concluded that in doing so, it erred by failing to take into account the full scope of the '029 patent. The '029 patent is not limited to compounds with a C_1-amide group, such as bimatoprost or the broader class of compounds described in the '819 patent. The scope of the independent claims of the '029 patent encompasses thousands of permutations of PGF analogs, including structures with all kinds of functional groups at the C_1 location, such as carboxylic acids, alkyl carboxylates, and hydroxyls. Apotex had the burden of showing that *any* compounds within the broad genus claimed by the '029 patent, including those that did not have C_1-amide groups, were obvious at the time of the invention.

The Federal Circuit found that the district court compounded its error by taking an overly cramped view of what the prior art teaches. The person of ordinary skill reading Johnstone would have found it replete with references to the advantages of using 17-phenyl PGF compounds generally, including those with carboxylic acids, esters, and other related groups at the C_1 location—all covered by the '029 patent's claims.

Johnstone discloses a preference for PGF analogs that had already been identified and were known in the art to have selective pharmacological activity, irrespective of whether such

---

[20] *KSR Int'l Co. v. Teleflex Inc.*, 550 U.S. 398, 406, 127 S. Ct. 1727, 167 L. Ed. 2d 705 (2007); Graham, 383 U.S. at 17-18.

[21] *Allergan*, 2013 U.S. Dist. LEXIS 13987, 2013 WL 286251, at *9.

[22] '819 patent, col. 3 ll. 9-18, col. 7 ll. 19-57.

activity was due to binding to the FP receptor or a variant, as well as whether there was a saturated or unsaturated bond at the C_5-C_6 location. Johnstone specifically taught that PGF analogs that were effective glaucoma drugs could grow hair. Specific classes of compositions of PGF analogs with specifically described structures and properties guided persons of ordinary skill in the art to compounds with similar structures that would fall within the scope of the '029 patent. "Obviousness based on structural similarity may be proven by the identification of some motivation that would have led one of ordinary skill in the art to select and modify a known compound in a particular way to achieve the claimed compound."[23] While success in employing the disclosed compounds to treat hair loss may not have been guaranteed, Johnstone's teaching provided sufficient guidance as to what parameters would lead to a reasonable expectation of success.

The district court's findings on secondary considerations failed to satisfy the requirement that "objective evidence of non-obviousness must be commensurate in scope with the claims which the evidence is offered to support."[24]

IV. Obviousness of the '404 Patent

The '404 patent emerged from the results of Allergan's clinical trials evaluating the safety and efficacy of bimatoprost eyedrops for glaucoma treatment, which were subsequently marketed as Lumigan®. At trial, Apotex identified four publications of clinical trials that they alleged disclose the ability of bimatoprost to promote eyelash hair growth (collectively the "Brandt references"), arguing that, among other grounds for invalidity, the Brandt references would render the '404 obvious in light of Johnstone.

The Brandt references include the following:

1. a presentation given by Dr. James Brandt, a clinical investigator working on Lumigan®, on October 23, 2000, disclosing three-month results from a bimatoprost eyedrop trial, which identifies the drug only by the name "Lumigan," stating that more than 5% of patients experience eyelash growth;
2. a press release issued by Allergan on October 23, 2000, containing the same information as the October 23, 2000 presentation;
3. a publication by Drs. Sherwood and Brandt in May 2001 disclosing that in a six-month glaucoma study, eyelash growth was reported in between 35 and 48% of patients receiving bimatoprost, depending on dose; and,

---

[23] *Altana Pharma AG v. Teva Pharms. USA, Inc..*, 566 F.3d 999, 1007 (Fed. Cir. 2009).

[24] *Application of Tiffin*, 448 F.2d 791, 792, 58 C.C.P.A. 1420 (CCPA 1971); *see also MeadWestVaco Corp. v. Rexam Beauty & Closures, Inc.*, 731 F.3d 1258, 1264-65 (Fed. Cir. 2013); *In re Huai-Hung Kao*, 639 F.3d at 1068; *In re Peterson*, 315 F.3d 1325, 1331 (Fed. Cir. 2003); *In re Hiniker Co.*, 150 F.3d 1362, 1369 (Fed. Cir. 1998).

4. a publication by Drs. Brandt, VanDenburgh (a named inventor of the '404 patent), Chen, and Whitcup in June 2001 disclosing that in a three-month study, eyelash growth was reported in 25.6 - 33.7% of patients receiving bimatoprost.

The first two Brandt references are indisputably prior art under § 102(b), as they were published more than one year before the priority date of the '404 patent, February 4, 2002. These references do not, however, expressly refer to bimatoprost. A person of ordinary skill hence would be unaware that the results reported therein were associated with bimatoprost. With respect to the more complete May and June 2001 references, the district court found that the '404 patent was invented prior to their publication date and, therefore, that the references were not prior art under § 102(a). The court based its conclusion on Allergan's evidence in support of its proposition that the date of invention predated the publication dates of the third and fourth publications.

On appeal, the Federal Circuit found that the evidence relied upon by the district court either did not relate to the claimed invention, or was uncorroborated, and thus insufficient to prove the earlier invention date.

In closing arguments at trial, Allergan argued that even if the '404 patent cannot claim an earlier priority date, the later Brandt references are still not § 102(a) art as they represent the work of the inventors themselves. "[O]ne's own work is not prior art under § 102(a) even though it has been disclosed to the public in a manner or form which otherwise would fall under § 102(a)."[25] Allergan claimed that the Brandt references were the product of the '404 patent's co-inventor Dr. VanDenburgh's work in designing and directing the Lumigan® clinical trials, relying predominately on Dr. Van-Denburgh's testimony regarding her role in the clinical trials, in which she described her role supervising and managing the work being done by various clinical study locations, including writing internal memoranda and reports. Thus, the Brandt references disclosed the work of only Dr. VanDenburgh, and was not prior art under § 102(a). On appeal, the Federal Circuit found that even if Allergan did not waive its right to bring this argument by failing to timely raise it at trial, the evidence did not support Allergan's position that the Brandt references disclosed the work of *only* inventors of the '404 patent, and that they were thus proper prior art under § 102(a).

The district court's findings on the obviousness of the '404 patent were limited to reiterating that the May and June 2001 Brandt references were not prior art (as a consequence of its finding on the mid-2000 invention date of the '404 patent) and that the '404 was not obvious in light of Johnstone alone.[26] The district court did make findings regarding inherent anticipation in light of the two earlier Brandt references.

On appeal, the Federal Circuit observed that Johnstone details at length how eyedrops containing latanoprost (marketed as the glaucoma drug Xalatan®) promote the eyelash hair growth through the mechanism of fluid containing latanoprost making topical contact with the eyelid and also discloses the topical application of latanoprost to treat hair loss. In light of

---

25 *In re Katz*, 687 F.2d 450, 454 (CCPA 1982).

26 *Allergan*, 2013 U.S. Dist. LEXIS 13987, 2013 WL 286251, at *10.

the Brandt reference's disclosure of bimatoprost's effect in growing eyelash hair, a person of ordinary skill in the art would have had substantial motivation to follow Johnstone and use topical application of bimatoprost to grow eyelash hair.

Likewise, the Brandt references provided a reasonable expectation of success for the topical application of bimatoprost. Clinical trials showed that nearly 50% of patients using bimatoprost in eyedrop form were experiencing eyelash hair growth. Johnstone additionally taught that fluid contacting the eyelid from eyedrops was the likely mechanism of hair growth. Accordingly, the Federal District reversed the district court's finding that the '404 patent is not invalid for reasons of obviousness.

Conclusion

For the foregoing reasons, the Federal Circuit reversed the district court's invalidity findings on the asserted claims of both the '029 and '404 patents, and vacated the court's injunction accordingly.

508983 Think like a Startup - Idea Protection

Abstract:

In today's economy, companies across industries are looking to capture the energy and zeal associated with the entrpreneurial spirit. Whether it's a newly formed company or a mature business, thinking like a startup can inject new life into an organization. But, with that newfound energy can sometimes come a lack of focus on certain details. This presentation will focus on how companies, young and old, can consider the creation and protection of intellectual property assets during the early period of starting a new business, pursuing new research, or developing a new business line. While companies often focus intently on their business plan, they may lose focus on the foundational aspects of that plan, i.e., the underlying intellectual property. Even if a company considers intellectual property protection, it may focus in on one type of intellectual property without considering how the different aspects of IP can be used together to develop a comprehensive plan. This presentation will introduce various aspects of Idea Protection and discuss how to develop an upfront plan of action to protect those assets.

519832 Clearing the Way: Using Post-Grant Proceedings to Invalidate Patents Blocking Your Freedom to Operate

Abstract:

Developing and marketing innovative products and processes is increasingly a capital-intensive undertaking with many technical challenges. More than ever, companies must ensure they have freedom to operate by analyzing whether any competitorâ€™s patents stand in the way and then finding ways to address such patents through validity challenges or licensing. This presentation will explore options in both the U.S. and Europe for challenging patents that could block your companyâ€™s ability to bring technology to market. Specifically, the speakers will discuss post-grant proceedings both in the U.S. Patent and Trademark Office and the European Patent Office that are powerful tools for invalidating patents, and also how companies can coordinate their U.S. and European challenges to achieve the best results.

AUTHOR INDEX